SPOTLIGHT ON ECOLOGY AND LIFE SCIENCE

PLANTS AND THEIR ENVIRONMENTS

ERIN LONG

PowerKiDS press

NEW YORK

Published in 2017 by The Rosen Publishing Group, Inc.
29 East 21st Street, New York, NY 10010

Editor: Theresa Morlock
Book Design: Michael Flynn
Interior Layout: Mickey Harmon

Photo Credits: Cover (leaf texture) PremiumVector/Shutterstock.com; cover (image) Yarygin/Shutterstock.com; p. 4 https://commons.wikimedia.org/wiki/File:Teofrasto_Orto_botanico_detail.jpg; p.5 Tetra Images/Getty Images; p. 7 Westend61/Getty Images; p. 9 (main) Evikka/Shutterstock.com; p. 9 (inset) https://upload.wikimedia.org/wikipedia/commons/2/2b/C._Darwin,_Insectivorous_Plants_Wellcome_L0031430.jpg; p. 11 K.Narloch/Shutterstock.com; p. 13 suwijaknook6644689/Shutterstock.com; p. 14 VisanuPhotoshop/Shutterstock.com; p. 15 Popperfoto/Contributor/Popperfoto/Getty Images; p. 15 (inset) https://commons.wikimedia.org/wiki/File:Peter_Rabbit_first_edition_1902a.jpg; p. 16 https://upload.wikimedia.org/wikipedia/commons/1/10/Leonardo_da_vinci%2C_Star_of_Bethlehem_and_other_plants.jpg; p. 17 KPG Payless2/Shutterstock.com; p. 19 Mr. SUTTIPON YAKHAM/Shutterstock.com; p. 21 sakarin nuchanong/Shutterstock.com; p. 22 Africa Studio/Shutterstock.com.

Cataloging-In-Publication Data

Names: Long, Erin.
Title: Plants and their environments / Erin Long.
Description: New York : PowerKids Press, 2017. | Series: Spotlight on ecology and life science| Includes index.
Identifiers: ISBN 9781499425871 (pbk.) | ISBN 9781499425901 (library bound) | ISBN 9781499425888 (6 pack)
Subjects: LCSH: Plants--Juvenile literature. | Plant ecology--Juvenile literature.
Classification: LCC QK49.L66 2017 | DDC 580--d23

Manufactured in the United States of America

CPSIA Compliance Information: Batch #BW17PK For further information contact Rosen Publishing, New York, New York at 1-800-237-9932.

CONTENTS

CHAPTER ONE

PEOPLE WHO STUDY PLANTS

Botany is the scientific study of plants, the way they work, and how they interact with their **environment**. People have always needed plants for food, shelter, medicine, and even oxygen. We couldn't live without them. Early human civilizations learned how to recognize plants, how to use them, and how to grow them. The first botanist may have been Theophrastus, a Greek philosopher who wrote about plants and their uses in about 300 BC.

The sciences of horticulture, agriculture, and forestry were developed from botany. Horticulturists **cultivate** plants for food and decoration. Agriculture, or farming, is how people work with plants and animals and adapt them to our needs. Forestry deals with the interactions between people and wildlife, plants, and **resources** in a forest **ecosystem**. People's relationship with plants and their environments continues to change with each new **generation**.

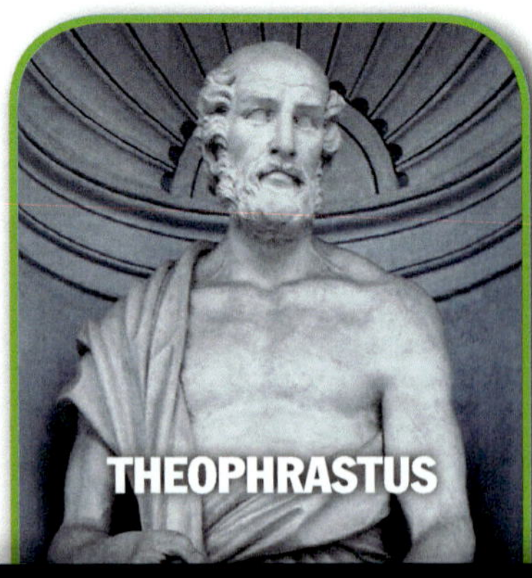

THEOPHRASTUS

Many plants can be used to treat illness or injury. Herbalists study the properties of plants and how they can be used as medicine.

CHAPTER TWO

CLASSIFYING PLANTS

There are over 400,000 species, or kinds, of plants that we know of, and scientists are still discovering more! They can be as small as a watermeal plant that's about 0.08 inch (2 mm) wide, or as tall as a 300-foot (91.4 m) giant sequoia. To keep track of all of these plants, scientists developed plant **classification**. In the plant kingdom, there are different ways to classify a plant, but the most simple way is to identify its genus and species. A genus is a group of species that share **traits** in common, while a species is a specific kind of **organism**.

One of the ways you can classify a plant is to study when it blooms. An annual plant goes through its entire life cycle in one season. After it dies, its seeds lie dormant, or not active, until they can grow next season. A biennial plant grows roots, stems, and leaves during one season, and then flowers, fruits, and seeds in a second season. Some examples of biennial plants are vegetables such as carrots and sugar beets. A perennial such as an apple tree survives for more than two years.

The study of plants and how to classify them is called plant taxonomy. Plant taxonomists work in colleges, botanical gardens, government offices, and many other places. They study and teach about the many plants on our planet.

CHAPTER THREE

ADAPTING TO THE ENVIRONMENT

The word "environment" is used to describe an organism's natural home, including other species and nonliving factors such as **climate**. These environments are part of biomes, some types of which are deserts, forests, and grasslands. Different plants have developed adaptations, or changes that help them to survive in their particular environments.

In the dry desert environment, succulent plants such as cacti are built to hold large amounts of water in their leaves, stem, and roots to be used as needed. In the rain forest, many trees have leaves with pointed drip tips, which are shaped so water runs off them quickly. Drip tips protect trees from fungi that would grow on them if they were soaked by heavy rainfall. Though a plant must adapt to survive in its environment, it also has the power to affect the climate and species in its community.

Charles Darwin studied the adaptations that insectivorous, or insect-eating, plants use to catch their food.

CHAPTER FOUR

PHOTOSYNTHESIS

Every plant needs a few key resources from its environment to keep it healthy. Instead of eating food, green plants turn energy from sunlight into the sugars that they need to survive. This process is called photosynthesis.

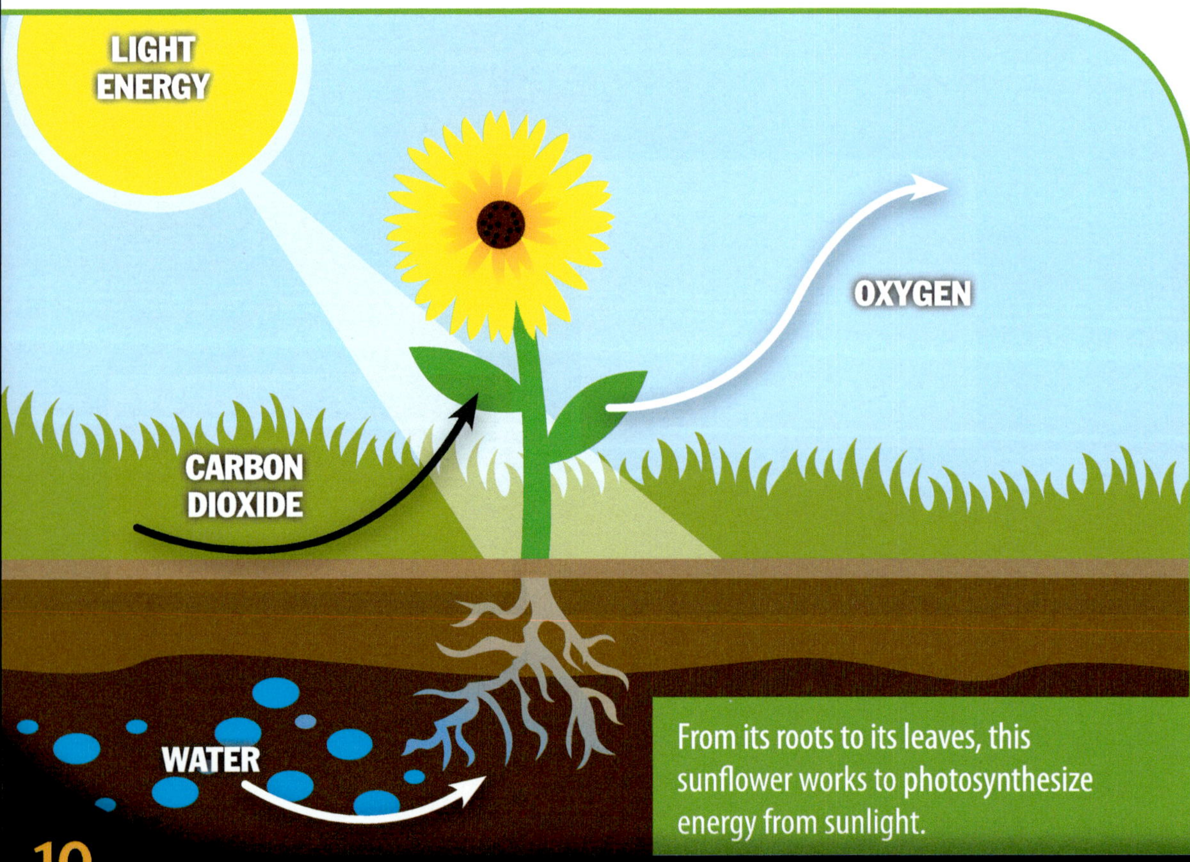

From its roots to its leaves, this sunflower works to photosynthesize energy from sunlight.

Chlorophyll provides the green color you see in plant leaves. It lets the leaves soak in the sunlight that they need for photosynthesis.

The three building blocks of photosynthesis are water, carbon dioxide, and sunlight. A plant uses its roots to collect water from the ground and passes it up to its stem and leaves. Carbon dioxide in the air enters the plant through little pores, or holes, in its leaves. The plant uses the water and carbon dioxide it collects to change the energy from sunlight into the sugars it needs to fuel its growth.

When animals and people breathe, they take in oxygen and release carbon dioxide. During the process of photosynthesis, plants release oxygen. This oxygen replaces what has been used by people and animals. Without the oxygen that plants produce, ecosystems would collapse.

CHAPTER FIVE

POLLINATION

Different plant species have different ways of producing **offspring**. For flowering plants, pollination is part of the process by which new seeds are grown. Flowering plants have a male part, called an anther, and a female part, called a stigma. The anther produces pollen, which must be moved to the stigma for the plant to make seeds. Flowering plants rely on pollinator animals, such as bees and hummingbirds, to carry the pollen for them.

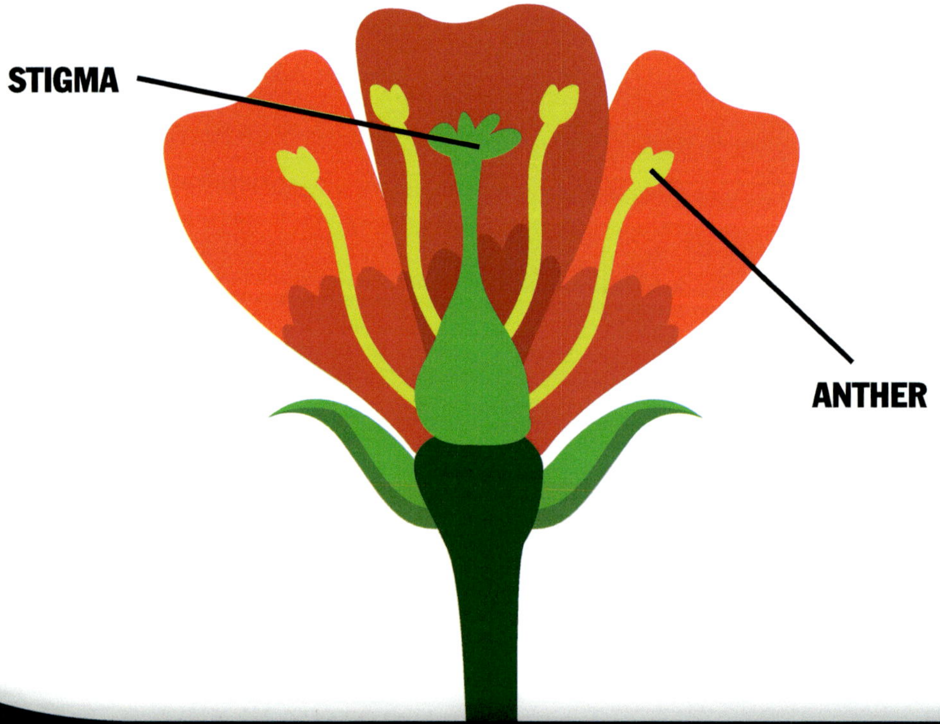

Butterflies are less effective pollinators than bees because their long legs don't pick up as much pollen.

Pollinators visit flowers to drink the nectar, or sweet juice, they produce. As they feed on the nectar, they brush against the flower's pollen, collecting it on their bodies. Pollinators deliver pollen to the stigma by flapping their wings to spread it on the wind, carrying it to other plants, or moving it to a different part of the same plant. When two organisms work together, it's called symbiosis. The process of pollination shows how plants and animals in an ecosystem can cooperate with each other to survive.

CHAPTER SIX

SPREADING ENERGY

In any ecosystem, energy is transferred from organism to organism through a food chain. In each food chain, there are three types of roles to fill: producer, consumer, and decomposer.

You've probably heard of the book *The Tale of Peter Rabbit* by Beatrix Potter. Potter also studied botany. She collected plants, made detailed drawings of them, and wrote down her observations.

Whether in a forest, desert, or grassland ecosystem, plants act as producers. An oak tree is able to make its own food through the process of photosynthesis, which means it doesn't need to rely on other living organisms for food. The oak tree is a producer because it makes, or produces, food for other members of its ecosystem. A deer that eats the acorns produced by the oak is a consumer. A wolf that eats the deer is also a consumer.

When producers and consumers die, their bodies are broken down by decomposers. Decomposers—such as mushrooms, earthworms, and bacteria—recycle the nutrients of dead organisms and restore them to the soil to encourage more growth.

CHAPTER SEVEN

THE IMPORTANCE OF PLANTS

Any ecosystem would collapse without plants. Plants provide shelter for other species in their communities. Trees are homes for animals such as woodpeckers, squirrels, chickadees, and many kinds of insects. Plants also affect the climate and landscape of their environment.

You probably know that Leonardo da Vinci was an artist and inventor, but did you know that he also studied plants and the way they worked? He observed that within a species, leaves grow in the same arrangement on each individual plant.

Practices such as logging, or cutting down trees, harm the environment by wiping out the plants that are needed to balance the ecosystem.

When plants absorb carbon dioxide during photosynthesis, it helps slow the process of **climate change** by reducing the amount of harmful gases in the atmosphere. Plants help keep the climate balanced by soaking up rainwater with their roots. Without plant life, flooding would be more common. Plants also fuel the water cycle through transpiration, or releasing water vapor into the atmosphere. Plant roots steady the soil by gripping it, preventing loose soil from being eroded, or worn away by wind or water.

CHAPTER EIGHT

HUMAN EFFECTS ON THE ENVIRONMENT

Of the known plant species, about one in five are at risk of extinction, or dying out. The greatest threats to the natural environment are from human activities. As the human **population** grows, plants face shrinking habitats, greater competition for resources, pollution, and shifts in climate conditions. Land development projects limit the amount of space for plants by taking away the resources they need to survive.

Pollution **contaminates** the environment with harmful materials. To protect crops from insects, large farms often use pesticides made of toxic, or poisonous, chemicals. The pesticides sink into the ground or are washed into streams and rivers. This pollutes nearby environments and can kill the organisms living there. Increasingly warmer temperatures change the natural conditions of the environment, which can wipe out plant species that are unable to adapt.

The beautiful lignum vitae has been endangered by loss of habitat due to construction projects.

CHAPTER NINE

HOW BIG IS YOUR FOOTPRINT?

When you pick your foot up off the ground, you can see the dent that it left behind on the grass. That's your footprint. When we talk about our ecological footprint, we're talking about the dent our lives make on the natural environment.

Our footprint is measured by how quickly we use Earth's resources and how much waste we leave behind. Some resources can't be replaced quickly enough to support our needs. At the most basic level, each person needs clothing, a place to live, food, and a way to get around. Everything people make and use spends resources and makes our footprint larger. People use plants to make food, medicine, buildings, furniture, and lots of other products, including toothpaste, paper, paint, and soap.

It's becoming more popular to practice organic, or all-natural, farming. Organic farms try to take care of plants they grow without using man-made chemicals. This can reduce an organic farmer's ecological footprint.

CHAPTER TEN

HELPING EARTH'S PLANTS

You can make choices every day that will help protect the environment. The best way to start is by thinking about ways you can conserve, or save, resources. When you reuse or recycle paper, not as much new paper needs to be made to replace it. Using less paper means that fewer trees will be cut down to make more. Think of all of the paper and cardboard products you use in a single day. Can you conserve those resources?

You can actively help the environment by planting your very own garden! You'll be able to see firsthand the strange and wonderful growth process of a plant. Every time you make a choice to conserve instead of spend resources, you are making your ecological footprint a little smaller.

GLOSSARY

classification (klaa-suh-fuh-KAY-shun) The act of organizing things into groups with shared traits.

climate (KLY-muht) The weather of an area.

climate change (KLY-muht CHANJ) The change in Earth's weather patterns due to higher levels of carbon dioxide in the atmosphere.

contaminate (kuhn-TAA-muh-nayt) To pollute or make something dirty or impure.

cultivate (KUHL-tuh-vayt) To raise crops by caring for the land and plants as they grow.

ecosystem (EE-koh-sis-tem) A natural community of living and nonliving things.

environment (in-VY-run-munt) The conditions that surround a living thing and affect the way it lives.

generation (jeh-nuh-RAY-shun) All of the people born and living at the same time.

offspring (OFF-spring) New life produced by an adult organism.

organism (OR-guh-nih-zum) An individual living thing.

population (pah-pyoo-LAY-shun) The number of organisms living in a place.

resource (RE-sors) Something that can be used.

trait (TRAYT) A quality that makes one person or thing different from another.

INDEX

PRIMARY SOURCE LIST

Page 9
Drosera rotundifolia. Illustration. From *Insectivorous Plants* by Charles Darwin. 1875. Published in London by John Murray.

Page 15
Beatrix Potter at Hill Top Farm, Sawrey. Photograph. ca. 1913. Now kept at Victoria and Albert Museum, London, England.

Page 16
Star of Bethlehem and other plants. Pen and ink over red chalk on paper. Leonardo da Vinci. ca. 1505. Now kept at the Royal Library, Windsor, England.

WEBSITES

Due to the changing nature of Internet links, PowerKids Press has developed an online list of websites related to the subject of this book. This site is updated regularly. Please use this link to access the list: www.powerkidslinks.com/sels/plant